HazCom Made Easier
*What You Need to Know About
Hazard Communication & GHS*

J. J. Keller
& Associates, Inc.®
Since 1953

Copyright 2017

J. J. Keller & Associates, Inc.
3003 Breezewood Lane, P.O. Box 368
Neenah Wisconsin 54957-0368
Phone: (800) 327-6868
Fax: (800) 727-7516
JJKeller.com
LCCN: 2011936240
ISBN: 978-1-60287-996-6
Canadian Goods and Services Tax (GST) Number:

Printed in the U.S.A.

EMPLOYEE'S RECEIPT

I acknowledge receipt of J. J. Keller's **HazCom Made Easier What You Need to Know About Hazard Communication & GHS**, which covers the following topics:

The Hazard Communication Standard

The Globally Harmonized System

Hazardous Chemical Classification

Safety Data Sheets

How to Read a Safety Data Sheet

How to Read a Container Label

Employee Information & Training

Working with Chemical Hazards

_____

EMPLOYEE'S NAME (PLEASE PRINT)

_____   _____

EMPLOYEE'S SIGNATURE   DATE

_____

COMPANY

_____

SUPERVISOR'S SIGNATURE

_____

NOTE: This receipt may be read and signed by the employee, countersigned by a company supervisor and placed in the employee's file.

Table of Contents

Table of Contents

Table of Contents

Due to the constantly changing nature of hazard communication, a website has been created to keep you informed of any critical updates affecting this *Handbook*. Go to **jjkeller.com/460h** for the latest information.

Reserved

Introduction

Over the years, commonplace chemicals like detergents, adhesives, pesticides, paints, and solvents have improved our lives, and made our work easier to do. However, chemicals can also be hazardous.

It is believed that one out of every four workers in the United States comes into contact with hazardous chemicals on the job.

Chemicals used in the workplace pose a wide range of health hazards such as irritation, sensitization, and carcinogenicity; or physical hazards such as flammability and corrosion. In fact, OSHA says that exposure to hazardous chemicals is one of the most serious threats facing American workers today.

Each year, hundreds of new chemicals are produced and almost 1.6 billion tons of chemicals are transported over our streets, highways, and railways.

The Occupational Safety and Health Administration (OSHA) estimates that 43 million workers produce or handle hazardous chemicals in more than five million workplaces across the country.

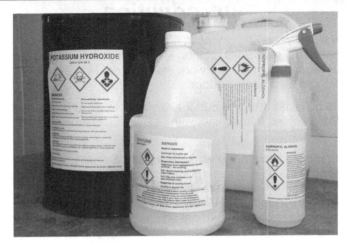

Hazard Communication Standard

OSHA's Hazard Communication Standard (HCS) is designed to ensure that information about chemical hazards and the ways that end users can work safely with the chemical is provided to downstream users.

This handbook will help you learn about the Standard, and how you can learn to be safe when working with chemical hazards in the workplace.

Optional Self-Check Quiz

At the end of each chapter you will find an optional self-check quiz which you can use to determine your level of comprehension and understanding of the materials presented.

What Is Hazard Communication?

In 1983 the Occupational Safety and Health Administration (OSHA) developed the **Hazard Communication Standard** (HCS), 29 CFR 1910.1200.

The HazCom standard is often called the "Right-to-Know" law or HazCom.

The HazCom standard is based on a simple idea — that you, as an employee, have both a right and a need to know:

- What chemicals you are exposed to,

- The hazards of working with those chemicals, and

- What steps you can take to protect yourself and those you work with.

Most chemicals used in the workplace have some hazard potential and so are covered by the rule.

> ➤ **To "use" a chemical means to package, handle, react, or transfer it.**

All employers with hazardous chemicals in their workplaces must:

- Compile a list of the hazardous chemicals known to be present (the chemical inventory).

- Obtain and provide employee access to Safety Data Sheets (SDSs).

- Ensure that all containers of hazardous chemicals are labeled.

- Conduct an effective training program for all potentially exposed employees.

- Prepare and implement a written hazard communication program.

What Is Hazard Communication?

Why Is a Standard Needed?

The HCS establishes uniform requirements to make sure that hazard information is transmitted to affected employers and exposed employees.

The areas covered by the HCS include:

- Hazard classification.
- A written hazard communication program.
- Labels and labeling.
- Safety data sheets (SDSs).
- Employee information and training.

OSHA believes that when employees understand the hazards of the chemicals that they work with, they will be more likely to take the steps necessary to protect themselves and their co-workers from those hazards.

> ➤ **OSHA has said that the revised HazCom standard now gives employees not only the "right-to-know" but also the "right-to-understand."**

What Is Hazard Communication?

As a result of "harmonizing" the HCS with the Globally Harmonized System (GHS) you will see changes in:

- **How chemical hazards are "classified."**

- **The look and content of container labels.**

- **The format and content of Safety Data Sheets (SDSs).**

- **HazCom information and training.** Employers are required to train workers on the new labels elements and safety data sheets format to facilitate recognition and understanding.

What Is the GHS?

In 2003, the United Nations (UN) adopted the **Globally Harmonized System of Classification and Labeling of Chemicals,** or GHS for short.

➤ **OSHA believes that revising the HCS to include the GHS will result in the safer handling of workplace chemicals and prevent over 500 workplace injuries and illnesses and 43 fatalities annually.**

What Is Hazard Communication?

The GHS includes criteria for the classification of health, physical and environmental hazards, as well as specifying what information should be included on labels of hazardous chemicals as well as safety data sheets.

Benefits of the GHS

Once fully implemented OSHA says that the HCS will:

- Improve hazard information in the workplace;
- Enhance worker understanding of hazards, especially for low and limited-literacy workers,
- Result in safer handling and use of chemicals; and
- Provide workers quicker and more efficient access to information on the safety data sheets.

The GHS revisions to the HCS standard for labeling and safety data sheets enable employees exposed to workplace chemicals to more quickly obtain and to more easily understand information about the hazards associated with those chemicals.

Employee _____

Instructor _____

Date _____

Location _____

What Is Hazard Communication Self-Check

1. The Hazard Communication Standard, 29 CFR 1910.1200, is also known as the " _____ - to- _____ " law.

2. The HazCom Standard gives you the right to all of the following, EXCEPT:

    a. The right to know what chemicals you are exposed to.

    b. The right to revise the written plan.

    c. The right to know the hazards of working with those chemicals.

    d. The right to know what steps you can take to protect yourself and those you work with.

3. GHS stands for the:

    a. Global Harmonization System

    b. Global Hazard Identification Scheme

    c. General Hazard Warning System

    d. Globally Harmonized System

4. To "use" a chemical means to package, handle, react, or transfer it.

    a. TRUE

    b. FALSE

5. OSHA believes that including GHS in HazCom will:

   a. Prevent 500 workplace injuries and illnesses annually.

   b. Prevent 43 workplace fatalities annually.

   c. Result in the safer handling of chemicals.

   d. All of the above.

Who Is Covered by the Standard?

OSHA's Hazard Communication Standard (HCS) applies to employers whose employees may be exposed to hazardous chemicals. It covers both general industry and construction employment.

The HCS applies to any hazardous chemical in the workplace that employees can be exposed to under normal conditions of use or in a **foreseeable emergency**.

> ➤ **"Foreseeable emergency" means any potential occurrence such as, but not limited to, equipment failure, rupture of containers, or failure of control equipment which could result in an uncontrolled release of a hazardous chemical into the workplace.**

Who Is Covered by the Standard?

Under OSHA's Hazard Communication Standard:

- Chemical manufacturers and importers must classify the hazards of the chemicals which they produce or import, and convey the hazard information to their downstream customers.

- Distributors are required to transmit chemical safety information to employers.

- Employers are required to provide information to their employees about the hazardous chemicals to which they are exposed.

Who Is Covered by the Standard?

Limited Provision of the HCS

In laboratories and workplaces where the chemicals are only handled in sealed containers, employers are not required to have written HazCom programs or a chemical inventory. They must, however, keep the labels on the containers that are received, keep the SDSs and make them available to employees, and train employees on the hazards of the chemicals in the work environment.

A "Right-to-Know"

As a worker, you need to know about the hazards of the chemicals you work with and what protective measures are available to prevent adverse effects from occurring.

OSHA incorporated portions of the Globally Harmonized System (GHS) into the revised HCS so that container labeling, SDS format and content, and chemical hazard determination are standardized and will look the same from workplace to workplace.

Who Is Covered by the Standard?

Studies have shown that the way the GHS presents hazard information is more easily understood by workers, and that workers retain the information longer.

Who Is Covered by the Standard?

Notes

Employee _____

Instructor _____

Date _____

Location _____

Who Is Covered by the Standard? Self-Check

1. The HazCom Standard covers:

    a. Labels and labeling

    b. Safety data sheets (SDSs)

    c. Employee information and training

    d. All of the above

2. The basic goal of the HCS is to be sure employers and employees know about work hazards and how to protect themselves.

    a. TRUE

    b. FALSE

3. Employers must provide information to employees about the hazardous chemicals to which they are

    _____ .

4. Chemical manufacturers and importers are required to do all of the following EXCEPT:

    a. Classify the hazards of the chemicals which they produce or import.

    b. Prepare container labels.

    c. Prepare safety data sheets.

    d. Create the workplace chemical inventory.

Who Is Covered by the Standard?

5. The HCS applies to workplaces where employees can be exposed to hazardous chemicals under normal conditions of use or in a _____ _____ .

What Makes a Chemical Hazardous?

OSHA' formal definition of a "hazardous chemical" is any chemical which is classified as a physical or health hazard, or which is a simple asphyxiant, combustible dust, pyrophoric gas, or hazard not otherwise classified (HNOC). In other words, any chemical that can hurt you.

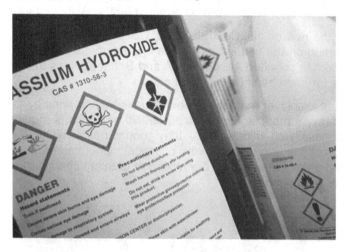

What Is Classification?

Chemical manufacturers and importers are required to evaluate the hazards of the chemicals they produce or import and then "classify" the chemical using the GHS criteria.

Under GHS, classification refers to identifying the hazard(s) of a chemical or mixture and then assigning it to one or more hazard classes using the defined GHS criteria.

What Makes a Chemical Hazardous?

Each hazard class can have up to 5 sub-categories. For instance, carcinogenicity has two categories:

- Category 1 is for known or presumed human carcinogens.

- Category 2 is for suspected human carcinogens.

By determining a chemical's hazard class and category, the hazard information that must be provided on the SDS and on the container label is pre-determined by the GHS system.

In the revised HazCom Standard, OSHA has included information on the general provisions for hazard classification, and has provided Appendixes A and B that address the criteria for each health or physical effect.

Your employer may rely upon the classification done by the chemical manufacturer or importer. No employer is required to classify chemicals unless they choose not to rely on the classification provided by the chemical manufacturer or importer or unless they are mixing or creating chemicals in the workplace.

What Makes a Chemical Hazardous?

Health Hazards

The HCS defines the following as health hazards, based upon the GHS classifications:

- Acute toxicity (any route of exposure)
- Skin corrosion or irritation
- Serious eye damage or eye irritation
- Respiratory or skin sensitization
- Germ cell mutagenicity
- Carcinogenicity
- Reproductive toxicity
- Specific target organ toxicity (single or repeated exposure)
- Aspiration hazard

Chemicals which are health hazards can cause illness right away (acute) or later on (chronic).

A rash that results from a one-time exposure of a chemical to the skin would be an **acute** health hazard.

Cancer that develops much later or is caused by repeated exposures to chemicals known to cause cancer would be a **chronic** health hazard.

Physical Hazards

Physical hazards refer to a chemical's physical properties, and mean that a material can easily burn, explode, or react violently when it comes in contact with another substance.

GHS uses the following classifications for physical hazards:

- Explosive
- Flammable (gas, aerosol, liquid, or solid)
- Oxidizer (liquid, solid, or gas)

What Makes a Chemical Hazardous?

- Self-reactive
- Pyrophoric (liquid or solid)
- Self-heating
- Organic peroxide
- Corrosive to metal
- Gas under pressure
- Emits flammable gas when in contact with water

Environmental Hazards

Environmental hazards refer to a chemical's ability to cause harm in the environment. GHS uses the following classifications for environmental hazards:

- Acute (short-term) aquatic hazards.
- Long-term aquatic hazards.
- Hazardous to the ozone layer.

What Makes a Chemical Hazardous?

While the GHS includes environmental classifications, OSHA did not adopt them into the HCS, since the Agency does not have authority over environmental issues.

Chemicals entering your workplace that were shipped from other countries which have adopted the GHS may have labels and SDSs with environmental hazards noted. That is why it is important for you to be familiar with them.

Other Hazardous Chemicals

OSHA has added specific definitions for other hazardous chemicals not included in the classifications found in the GHS:

- **Pyrophoric gases:** Labels for pyrophoric gases must include the signal word "danger" and the hazard statement "catches fire spontaneously if exposed to air."

- **Simple asphyxiants:** Labels for simple asphyxiants must include the signal word "warning" and the hazard statement "may displace oxygen and cause rapid suffocation."

- **Combustible dust:**Labels for combustible dust must include the signal word "warning" and the hazard statement "May form combustible dust concentrations in the air."

This information must also appear on the SDS.

Hazard Not Otherwise Classified

There are chemicals for which there is evidence of adverse physical or health effects, but which do not meet the specified criteria for any of the physical or health hazard classifications. These chemicals are referred to as a "Hazards Not Otherwise Classified (HNOC)."

Classification as an HNOC does not mean the chemical poses no hazards - only that it does not fit into one of the established GHS hazard classes, or that it falls into a hazard category that OSHA has not adopted, such as Acute Toxicity - Category 5.

What Makes a Chemical Hazardous?

The SDS must identify the chemical as an HNOC.

What About Biological Substances?

While infectious biological substances are hazardous, they are not regulated by the Hazcom Standard.

Consumer Product Exemption

Consumer products, as defined in the Consumer Product Safety Act (15 USC 2051 et seq.), are exempt from coverage under the HazCom standard if:

- It is used in the workplace for the purpose intended by the manufacturer or importer, and if

- It's use results in a duration and frequency of exposure which is not greater than the range of exposures that could be reasonably experienced by consumers when used for the purpose intended.

What Makes a Chemical Hazardous?

Articles

Articles, as defined in the standard, are also exempt. Articles are manufactured items other than a fluid or a particle which:

- Are formed to a specific shape or design during manufacturer,

- Have end use function(s) dependent in whole or in part upon its shape or design during end use,

- Under normal conditions of use do not release more than very small quantities, minute or trace amounts of a hazardous chemical, and

- Do not pose a physical hazard or health risk to employees.

An example of an article might be a brick or a galvanized pipe. By themselves, they do not give off any hazardous chemicals, but if you cut the brick or weld the pipe, you create hazards, so they would no longer be considered articles.

Notes

Employee _____

Instructor _____

Date _____

Location _____

What Makes a Chemical Hazardous? Self-Check

1. OSHA defines a "hazardous chemical" as any chemical which is:

    a. Classified as a physical or a health hazard.

    b. A simple asphyxiant, combustible dust, or pyrophoric gas.

    c. A chemical which is a hazard not otherwise classified (HNOC).

    d. All of the above.

2. All of the following are classified as health hazards EXCEPT:

    a. Eye irritants.

    b. Organic peroxides.

    c. Acute toxins.

    d. Aspiration hazards.

3. All of the following are classified as physical hazards EXCEPT:

    a. Explosives.

    b. Oxidizers.

    c. Pyrophorics.

    d. Carcinogens.

4. OSHA has adopted environmental hazards into the HCS, even though the agency has no authority over environment issues.

    a. TRUE

    b. FALSE

5. A "hazard not otherwise classified" (HNOC) is a chemical for which there is evidence of adverse physical or health effects, but which do not meet the specified criteria for any of the physical or health hazard classifications.

    a. TRUE

    b. FALSE

What Must My Employer Do?

The HCS requires chemical hazard information to be pre-pared and transmitted to all downstream users of those chemicals.

Both your employer and the chemical manufacturer or importer have responsibilities to supply you with that information.

Manufacturers and Suppliers

Chemical manufacturers and importers must evaluate the hazards of the chemicals they produce or import, label the containers, and provide safety data sheets (SDSs) to the employers that receive those chemicals.

Employers do not have to reclassify any of the hazardous chemicals they receive. They can rely upon the information provided by the chemical manufacturer or importer.

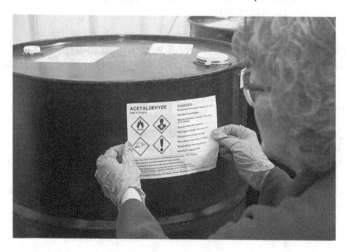

What Must My Employer Do?

Employers

The HCS requires your employer to:

- Identify all hazardous materials in the workplace.

- Develop a hazard communication program, including a hazardous chemical inventory.

- Get an SDS for each hazardous chemical in the workplace.

- Label all hazardous materials containers in the workplace.

- Design and implement an employee protection program.

- Train employees about the standard, and provide information to employees on how they can protect themselves from chemical hazards in the workplace.

- Assure employee access to SDSs and to the company written HazCom program.

Employers that "use" hazardous chemicals must have a program to ensure the information is provided to exposed employees. "Use" means to package, handle, react, or transfer.

State Plan States

Some states have different requirements than the federal Standard, however, states may not have requirements which are less effective than federal OSHA. If your employer is in an OSHA-approved State Plan State, then your workplace must comply with the State's requirements, which might be different than OSHA's.

What Must My Employer Do?

Employee _____

Instructor _____

Date _____

Location _____

What Must My Employer Do? Self-Check

1. The HCS requires information to be prepared and transmitted regarding all hazardous chemicals.

   a. TRUE

   b. FALSE

2. The HCS requires your employer to do all of the following EXCEPT:

   a. Identify all hazardous materials in the workplace.

   b. Get an SDS for each hazardous chemical in the workplace.

   c. Train employees about the standard.

   d. Provide employees with a yearly toxicological screen.

3. According to OSHA, to "Use" means to:

   a. Package it.

   b. Handle it.

   c. Transfer it.

   d. All of the above

4. Employers must ALWAYS reclassify hazardous chemicals which arrive in the workplace.

   a. TRUE

   b. FALSE

5. States which have different workplace safety requirements than the federal Standard are called
   _____ _____ _____ .

The Written Program

All workplaces where employees are exposed to hazardous chemicals or could be exposed in a foreseeable emergency must have a written HazCom plan that describes how the company will meet its obligations under the standard.

The written plan must show how your company has met all of the standard's requirements.

What Is in the Written Program?

The written program must explain:

- How safety data sheets are received and kept.
- The names of those in charge of labeling, SDS collection, and inventory.

- How hazardous chemical training is done. This must include:

  o Methods used for chemical-specific training,

  o The type of safety precautions taught,

  o Emergency and first aid training provided, and

  o Types of training provided for certain non-routine tasks, such as cleaning reactor vessels.

- The hazards associated with unlabeled pipes.

- How this information and training will be provided to contract workers.

The written program will also contain a copy of the chemical inventory.

Hazardous Chemical Inventory

To protect yourself from hazardous chemicals you must first know what chemicals are in the workplace. This is why your employer is required to maintain a list of hazardous chemicals called the Chemical Inventory.

The inventory is part of the written hazard communication program, and will eventually serve as the list of all chemicals for which there must be an SDS available.

This inventory must include:

- All hazardous chemicals used in your company,

- Hazardous chemicals taken off-site for use in other locations, and

- Hazardous chemicals stored away from the main facility.

The list can be compiled for the workplace as a whole or for individual work areas.

The chemical inventory should reference chemicals using the name found on the container label and on the SDS. All three — the chemical inventory, the container label, and the SDS — should match. The intent is that in an emergency you can look at the label and quickly locate the correct SDS.

Preparing this list may help your employer to determine where less hazardous chemicals can be substituted, or if chemicals which are not being used anymore can be eliminated from the inventory altogether.

The list also lets you know what chemicals must have SDSs.

Your Responsibilities

You have a right to see the chemical inventory for your establishment. Even though your employer is responsible for maintaining the hazardous chemical inventory, employees are the ones who work near the chemicals and who should be aware of the risks involved.

Therefore, it is to your advantage to help keep the inventory current.

If you notice a new hazardous chemical in your work area, check the inventory to see if it's been added. If the chemical isn't listed, let your supervisor know.

The Written Program

Multi-Employer Workplaces

The written plan will explain how HazCom information will be provided to other employer's employees in multi-employer workplace settings.

If the workplace has multiple employers onsite (for example, a construction site), the rule requires these employers to ensure that information regarding hazards and protective measures be made available to the other employers onsite, where appropriate.

How Can I Get a Copy of the Written Program?

Your employer must make copies of the written program available to you or to your representative, such as a union official.

Usually you will receive or be told where to get a copy of the written program during hazard communication training. You may always ask your supervisor for a copy.

Employee _____

Instructor _____

Date _____

Location _____

The Written Program Self-Check

1. The HazCom Written Plan describes how the company will meet its obligations under the HazCom Standard.

   a. TRUE

   b. FALSE

2. The written program must do all of the following EXCEPT:

   a. Explain how safety data sheets are received and kept.

   b. Include information on HazCom training.

   c. Be available to employees electronically.

   d. Include a copy of the chemical inventory.

3. The chemical inventory serves as the list of all chemicals for which there must be an _____ available.

4. Your employer must make copies of the written program available to your or to your representative.

   a. TRUE

   b. FALSE

5. The hazardous chemical inventory must include all of the following EXCEPT:

   a. All hazardous chemicals used at your facility.

   b. Hazardous chemicals taken offsite for use in other locations.

   c. Hazardous chemicals stored away from the main facility.

   d. All hazardous wastes generated at the facility.

Hazard Warning Labels

Chemical hazard warning labels are one way of informing you of hazards and how to protect yourself when using or storing that material.

Hazard warning labels must be placed on every container of hazardous chemicals which is in the workplace.

Labels and other forms of warning must be legible, in English, and prominently displayed on the container or readily available in the work area throughout each shift.

Employers may add information in languages other than English as long as it is presented in English as well.

A hazard warning label is not intended to be the sole or most complete source of hazard information. For complete hazard information, always refer to the Safety Data Sheet (SDS).

What Must Be on the Label?

Containers of hazardous chemicals which leave the workplace, referred to as "shipped containers," must be labeled with these six required elements:

- Product identifier

- Pictogram

- Signal word

- Hazard statement(s)

- Precautionary statement(s)

- Name, address, and telephone number of the chemical manufacturer, importer, or other responsible party

Your employer may use workplace or in-house labeling in place of GHS-style labeling on containers that do not leave the workplace. This alternative labeling must include:

- Product identifier, and

- Words, pictures, symbols, or combinations thereof, which provide information regarding the hazards of the chemicals.

Is There a Specified Label Format?

Neither OSHA nor GHS specify any label format or label design. The pictograms, signal word, and hazard statements should be located together on the label.

What Is the Product Identifier?

The product identifier is the name or number used on the label and SDS. It can be a chemical name, a product name, or some other unique identifier that allows you to locate the SDS quickly.

What Are Pictograms?

For HCS purposes, a pictogram is a symbol on a white background with a red border that is intended to convey specific information about the hazards of a chemical. For in-plant or workplace labeling only, OSHA says that pictograms may have a black border, rather than a red border.

The pictograms which appear on the label are determined by the chemical's hazard classification.

Flame over circle is used to denote oxidizers.

Flame is used to mark flammables, self-reactives, pyrophorics, self-heating substances, substances that emit flammable gases, and organic peroxides.

Exploding bomb is used to signify explosives, self-reactives, and organic peroxides.

Skull and crossbones identifies acutely toxic (severe) or fatal substances.

Corrosion is used to signify corrosives, substances that are skin corrosives or can cause skin burns, or substances that can damage the eye.

Gas cylinder indicates gases under pressure.

Health hazard is to denote carcinogens, respiratory sensitizers, reproductive toxins, target organ toxins, mutagens, and aspiration toxins.

Exclamation mark is used to signify irritants, skin sensitizers, acute toxins (harmful), chemicals with narcotic effects, and respiratory tract irritants. While not mandatory under OSHA it will also identify substances hazardous to the ozone.

Environmental denotes substances that are environmental toxins. While OSHA does not address environmental hazards under the HCS, you may see this pictogram on labels.

Pictogram   Product identifier   Signal word

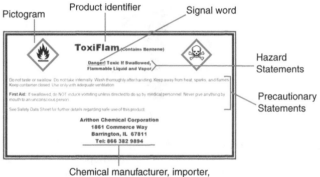

Hazard Statements

Precautionary Statements

Chemical manufacturer, importer, or other responsible party.

What Is a Signal Word?

The signal word is used to alert the user to a potential hazard and is determined by the hazard class and category of the chemical.

When a signal word is required on a label, HazCom requires that it be either:

- **"Danger"** for more severe hazards, and

- **"Warning"** for less severe ones.

When the signal word "Danger" appears, the signal word "Warning" shall not appear.

What Are Hazard Statements?

Hazard statements are standardized phrases assigned to a specific hazard class and category.

Hazard statements are used to describe the nature of the hazard(s), including, where appropriate, the degree of hazard.

Examples are "Causes serious eye damage." and "Fatal if swallowed."

What Are Precautionary Statements?

Precautionary statements are standardized phrases assigned to a hazard class and category.

There are four types of precautionary statements, covering prevention, response in case of accidental spills or exposure, storage, and disposal.

Use of precautionary statements, while optional in the GHS system, are mandatory in OSHA's HazCom standard.

The statements are linked to each hazard class and category. Examples are "Store locked up." and "Wear protective gloves/protective clothing."

Supplier Identification

Supplier identification refers to the name, address, and telephone number of the chemical manufacturer, importer, or other responsible party.

Hazard Warning Labels

Workplace Labels

OSHA allows employers to use any type of warning labeling in the workplace, as long as that label has:

- The name of the material, and

- Information about the health and physical hazards.

The employer can determine what will work best for the facility as long as employees have been trained on both the GHS-style labels and the in-house labeling system.

Read the Label — Every Time!

Always look at the label, every time you use a hazardous chemical. Conditions may change or new ways of protecting workers from hazards may be found. That new information would be on the label.

What if I Have Questions About the Label?

If you are not sure what the label is telling you, do not use the chemical until your questions are answered. Check the SDS or ask your supervisor for help.

You should not work with any hazard chemical if you have any safety questions or are unsure of the hazards.

What if the Container Doesn't Have a Label?

All containers of hazardous materials in the workplace must be properly labeled.

> **Don't open the container or use the chemical if you are not sure what it is or what the hazards of the material are.**

If there are containers which are not labeled, or with labels that you cannot read or that are torn or partially missing, report it to your supervisor immediately.

Exemptions to Labeling

In-plant labeling exemptions include:

- Signs or placards for stationary containers in work areas that have similar contents and hazards.

- Operating procedures, process sheets, batch tickets, blend tickets, and similar written materials on stationary process equipment.

- Portable containers.

- Pipes or piping systems.

What Is a "Portable Container?"

A portable container is a container used to transfer a hazardous chemical from a labeled container and is intended only for the immediate use of the employee who performs the transfer.

Hazard Warning Labels

OSHA says that "immediate use" means the chemical in the portable container can only be used by the employee who transfers it from a labeled container and must be used on that work shift.

Any remaining chemical in an unlabeled portable container **may not** be passed along to another employee to use.

Portable containers do not have to be labeled, **unless** the container and its contents are passed along to another employee.

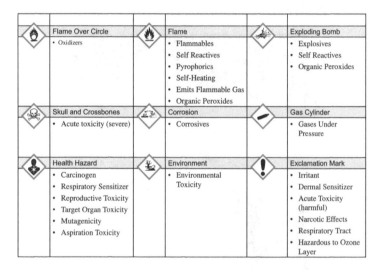

	Flame Over Circle		Flame		Exploding Bomb
	• Oxidizers		• Flammables • Self Reactives • Pyrophorics • Self-Heating • Emits Flammable Gas • Organic Peroxides		• Explosives • Self Reactives • Organic Peroxides
	Skull and Crossbones		Corrosion		Gas Cylinder
	• Acute toxicity (severe)		• Corrosives		• Gases Under Pressure
	Health Hazard		Environment		Exclamation Mark
	• Carcinogen • Respiratory Sensitizer • Reproductive Toxicity • Target Organ Toxicity • Mutagenicity • Aspiration Toxicity		• Environmental Toxicity		• Irritant • Dermal Sensitizer • Acute Toxicity (harmful) • Narcotic Effects • Respiratory Tract • Hazardous to Ozone Layer

44

Employee _____

Instructor _____

Date _____

Location _____

Hazard Warning Labels Self-Check

1. Hazard warning labels must be placed on each container of hazardous chemical which is in the workplace.

   a. TRUE

   b. FALSE

2. OSHA's required container label elements include all of the following EXCEPT:

   a. EPA hazard warnings.

   b. Product identifier.

   c. Signal words.

   d. Pictograms.

3. Container labels must be printed in English, Spanish, and French.

   a. TRUE

   b. FALSE

4. On in-plant labeling, pictograms must have a red border.

   a. TRUE

   b. FALSE

5. The _____ _____ _____ is not intended to be the sole or most complete source of hazard information.

6. OSHA requires that the pictogram, product identifier, and signal word be located together on the label.

   a. TRUE

   b. FALSE

7. The _____ and _____ pictogram denotes acutely toxic substances.

8. Don't open the container or use a chemical if you are not sure what is in the container or what the hazards of the material are.

   a. TRUE

   b. FALSE

9. Of the two signal words "DANGER" and "WARNING," _____ is used for more severe hazards.

Safety Data Sheets

OSHA requires that employers provide employees with access to an SDS for each hazardous chemical they use.

OSHA says that SDSs must be:

- Printed in English.

- Readily accessible in each work area during each work shift.

SDSs are the primary source for hazard information for the chemicals in your workplace.

Safety Data Sheets

Why Should I Look at the SDS?

The SDS not only helps you determine what safety measures are needed, but it could also save valuable time in the event of an emergency. You should consult the SDS to:

- Verify the information on the container label regarding safety and hazards.

- Make sure you are using the correct PPE for the chemical.

- Determine if any symptoms you might be experiencing can be attributed to the chemical.

SDS Number: **UR-101** * * * * * *Effective Date: 05/27/12* * * * * * Supercedes: **05/08/00**

Solvent X (contains Methylene Chloride)

1. Product and Company Identification

Product Name: Solvent X
Synonyms: Machine solvent, industrial solvent, equipment flushing solvent
CAS No.: Not applicable to mixtures
Molecular Weight: Not applicable to mixtures
Emergency Contact: +1-123-555-1212

Product Use: Various industrial uses
Restrictions: Not to be used as a skin cleaner
Chemical Formula: Not applicable to mixtures
Product Codes: UR-101
Manufacturer: U. R. Chemical
Little Hill Dr.
Little Mill, KS 01234
Telephone +1-123-555-0101
Fax +1-123-555-0102

2. Hazards Identification

GHS CLASSIFICATION:

	Health		Environmental	Physical
Acute Toxicity:	Category 4	Acute Toxicity:	None Known	
Skin Irritation:	Category 2	Chronic Toxicity:	None Known	
Skin Sensitization:	NO			
Eye:	Category 2B			

GHS LABEL:	⬥ ⬥	Signal Word: Warning	HMIS CLASSIFICATION: CLASS D, DIVISION 1

Hazard Statements	Precautionary Statements
H320 Causes eye irritation	P210 Keep away from heat/sparks/open flames/hot surfaces - No smoking
H335 May cause respiratory irritation	P261 Avoid breathing dust/fume/gas/mist/vapors/spray
H336 May cause drowsiness or dizziness	P280 Wear protective gloves/protective clothing/eye protection/face protection
H351 Suspected of causing cancer	P271+P313 Get medical advice/attention
	P403+P233 Store in a well-ventilated place. Keep container tightly closed
	P501 Dispose of contents/container in accordance with local regulation

3. Composition/Information on Ingredients

Ingredient	CAS No	Percent	Hazardous
Methylene Chloride	75-09-2	80%	Yes
Ethyl Alcohol	64-17-5	30%	Yes
Acetic Acid	64-19-7	1%	Yes
Paraffinic Petroleum Distillates	64742-65-0	1%	Yes

4. First Aid Measures

Inhalation: Remove to fresh air. Get medical attention for any breathing difficulty. If breathing stops, administer artificial respiration.
Ingestion: Do not induce vomiting. Keep individual calm. Obtain medical attention immediately.
Skin Contact: Remove contaminated clothing and shoes. Wash exposed area with soap and water for at least 15 minutes. Get medical advice if irritation develops. Wash contaminated clothing before reuse.
Eye Contact: Flush thoroughly with running water for at least 15 minutes, occasionally lifting the eyelids. Obtain medical attention.

5. Fire Fighting Measures

Fire: Fire is possible if concentrated vapors are exposed to elevated temperatures.
Fire Extinguishing Media: Water spray, dry chemical, alcohol foam, or carbon dioxide
Special Information: In the event of a fire, wear full protective clothing and NIOSH-approved self-contained breathing apparatus with full facepiece operated in the pressure demand or other positive pressure mode.

6. Accidental Release Measures

Ventilate area of leak or spill. Wear appropriate personal protective equipment as specified in Section 8. Spills: Pick up spill for recovery or disposal and place in a closed container. Keep spilled material away from sewage/drainage systems and waterways.

7. Handling and Storage

Keep in a tightly closed container, stored in a cool, dry, ventilated area. Protect against physical damage. Isolate from incompatible substances. Observe all warnings and precautions listed for the product.

8. Exposure Controls/Personal Protection

Airborne Exposure Limits:
OSHA Permissible Exposure Limit (PEL):
methylene chloride = 5000 ppm (TWA);
ethyl alcohol = 1000 ppm;
acetic acid = none established;
paraffinic petroleum distillates = 400 ppm.
ACGIH Threshold Limit Value (TLV):
methylene chloride = 50 ppm (TWA);
ethyl alcohol = 1000 ppm;
acetic acid = none established;
paraffinic petroleum distillates = 400 ppm.

Ventilation System:
Sufficient to maintain vapor concentrations below TLV. Do not use in a closed or confined space.
Personal Respirators (NIOSH Approved):
If the exposure limit is exceeded, wear an appropriate NIOSH N95 respirator, full-facepiece respirator, or airfeed hood.
Skin Protection:
Wear impervious protective clothing, including gloves and apron, to prevent skin contact.
Eye Protection:
Use chemical safety goggles or full face shield when splashing is a concern. Maintain eye-wash fountain and quick-drench facility in the work area.
Other Control Measures:
Protective equipment for laboratory bench use should be chosen using professional judgment based on the size and type of reaction.

Notice: Chemical and other data represented in this safety data sheet image is for example purposes only.

How to Read an SDS

Under the requirements of the GHS, each SDS must have the headings listed below in the order listed.

Section 1. **Identification** — This section identifies the chemical on the SDS as well as the recommended uses. It also provides the essential contact information of the supplier. The required information consists of:

      o Product identifier used on the label and any other common names or synonyms by which the substance is known.

      o Name, address, phone number of the manufacturer, importer, or other responsible party, and emergency phone number.

      o Recommended use of the chemical and any restrictions on use.

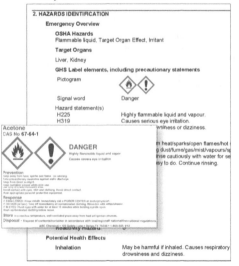

Section 2.   **Hazard(s) identification** — Identifies the hazards of the chemical and the appropriate warning information associated with those hazards. The required information consists of:

    o   The hazard classification of the chemical (e.g., flammable liquid, category).

    o   Signal word.

    o   Hazard statement(s).

    o   Pictogram(s).

    o   Precautionary statement(s).

    o   Description of any hazards not otherwise classified.

    o   For a mixture that contains an ingredient(s) with unknown toxicity, a statement describing how much (percentage) of the mixture consists of ingredient(s) with unknown acute toxicity. Please note that this is a total percentage of the mixture and not tied to the individual ingredient(s).

Section 3.   **Composition/information on ingredients** — Identifies the ingredient(s) contained in the product, including impurities and stabilizing additives. This section includes information on substances, mixtures, and all chemicals where a trade secret is claimed. The required information consists of:

Substances

    o   Chemical name.

    o   Common name and synonyms.

    o   Chemical Abstracts Service (CAS) number and other unique identifiers.

○ Impurities and stabilizing additives, which are themselves classified and which contribute to the classification of the chemical.

Mixtures

○ The chemical name and concentration (i.e., exact percentage) of all ingredients which are classified as health hazards and are:

  ○ Present above their cut-off/ concentration limits or

  ○ Present a health risk below the cut-off/ concentration limits.

○ The concentration (exact percentages) of each ingredient must be specified except concentration ranges may be used in the following situations:

  ○ A trade secret claim is made,

  ○ There is batch-to-batch variation, or

  ○ The SDS is used for a group of substantially similar mixtures.

Chemicals where a trade secret is claimed

○ A statement that the specific chemical identity and/or exact percentage (concentration) of composition has been withheld as a trade secret is required.

Section 4. First-aid measures — Initial care that should be given to an individual who has been exposed to the chemical. The required information consists of:

○ Necessary first-aid instructions by relevant routes of exposure (inhalation, skin and eye contact, and ingestion).

○ Description of the most important symptoms or effects, and any symptoms that are acute or delayed.

○ Recommendations for immediate medical care and special treatment needed, when necessary.

Section 5. **Firefighting measures** — Recommendations for fighting a fire caused by the chemical. The required information consists of:

○ Suitable extinguishing equipment.

○ Specific hazards that develop from the chemical during the fire.

○ Recommendations on special protective equipment or precautions for firefighters.

Section 6. **Accidental release measures** — Recommendations on the appropriate response to spills, leaks, or releases, including containment and cleanup practices and recommendations for:

○ Precautions and PPE.

○ Emergency procedures, including instructions for evacuations, and appropriate protective clothing.

○ Methods and materials used for containment.

○ Cleanup procedures.

Section 7. **Handling and storage** — Guidance on the safe handling practices and conditions for safe storage of chemicals. The required information consists of:

○ Recommendations for handling incompatible chemicals, minimizing the release of the chemical into the environment, and providing advice on general hygiene practices.

SAFETY DATA SHEETS

53

       ○   Recommendations on the conditions for safe storage, including any incompatibilities. Provide advice on specific storage requirements.

Section 8. **Exposure controls/personal protection** — Indicates the exposure limits, engineering controls, and personal protective measures that can be used to minimize worker exposure, consisting of:

       ○   OSHA Permissible Exposure Limits (PELs), ACGIH Threshold Limit Values (TLVs), and any other exposure limit used or recommended by the chemical manufacturer, importer, or employer preparing the safety data sheet, where available.

       ○   Appropriate engineering controls (e.g., use local exhaust ventilation, or use only in an enclosed system).

       ○   Recommendations for personal protective measures and PPE.

       ○   Any special requirements for PPE, protective clothing or respirators (e.g., type of glove material, such as PVC or nitrile rubber gloves; and breakthrough time of the glove material).

9. Physical and Chemical Properties

Appearance:
Clear, yellowish liquid.
Odor:
Sweet, aromatic odor.
Solubility:
1.4 grams per 100 grams at 77°F
Specific Gravity:
1.2 at 77°F
pH:
No information found.
% Volatiles by volume @ 21°C (70°F):
Not available
Boiling Point:
No information found.

Melting Point:
199°C (390°F) Decomposes.
Freezing Point:
-96.7°C (-142.1°F)
Flash Point:
16°C (61°F)
Flammability:
Non-flammable liquid. Vapor will burn at high temperatures.
Vapor Density (Air=1):
No information found.
Vapor Pressure (mm Hg):
No information found.
Evaporation Rate (BuAc=1):
No information found.

10. Stability and Reactivity

Stability:
Stable under ordinary conditions of use and storage.
Hazardous Decomposition Products:
Burning may produce carbon monoxide, carbon dioxide, nitrogen oxides.

Hazardous Polymerization:
Will not occur.
Incompatibilities:
Strong oxidizers and reactive metals.
Conditions to Avoids:
Incompatibles and hot surfaces which can cause thermal decomposition.

11. Toxicological Information

Methylene chloride: oral rat LD50: 16000mg/kg; inhalation rat LC50: 52gm/m3; investigated as a tumorigen, mutagen, reproductive effector.
Ethyl alcohol: oral rat LD50= 7060mg/kg; inhalation rat LC50= 20,000ppm/10H; investigated as a tumorigen, mutagen, reproductive effector.
Acetic acid: No LD50/LC50 information found relating to normal routes of occupational exposure.
Paraffinic petroleum distillates: Not known.

12. Ecological Information

Environmental Fate:
When released into the soil, this material may biodegrade to a moderate extent. When released into the soil, this material is expected to leach into groundwater. When released into water, this material may biodegrade to a moderate extent. When released into the air, this material may be moderately degraded by reaction with photo-chemically produced hydroxyl radicals. When released into air, this material is expected to have a half-life between 10 and 30 days.

Ecotoxicological Data:
96 hr. NOEL (minnow) 110 mg/L.
24 hr EC50 (Daphnia Magna) 460 mg/L.

13. Disposal Considerations

Recover or recycle if possible. It is the responsibility of the waste generator to determine the proper waste classification and disposal methods in compliance with applicable federal, state, and local regulations. Do not dispose of in the environment, in drains or water courses. Drain container thoroughly. Send to recycler/reclaimer.

14. Transport Information

Transport in accordance with all federal, state, and local regulations.
UN number 2929
UN proper shipping name   Toxic liquid, flammable, n.o.s.
Hazards class 6.1; Packing Group III

15. Regulatory Information

----------------------------\Chemical Inventory Status - Part 1\----------------------------

Ingredient	TSCA	EC	Japan	Australia
Methylene Chloride (75-09-2)	Yes	Yes	Yes	Yes
Ethyl Alcohol (64-17-5)	Yes	Yes	Yes	Yes
Acetic Acid (64-19-7)	Yes	Yes	Yes	Yes
Paraffinic Petroleum Distillates (64742-65-0)	Yes	Yes	No	Yes

Chemical Weapons Convention: No   TSCA 12(b): Yes   CDTA: No
SARA 311/312: Acute: Yes   Chronic: Yes Fire: No Pressure: No
Reactivity: No   (Mixture / Liquid)

Australian Hazchem Code: None allocated.
Poison Schedule: None allocated.

16. Other Information

NFPA Ratings: Health: 3 Flammability: 1 Reactivity: 0
HMIS Ratings: Health: 3 Flammability: 1 Physical Hazard: 0
This SDS was prepared by U. R. Chemical.

Revision Information: Revised 05/27/12

Notice: Chemical and other data represented in this safety data sheet image is for example purposes only.

Section 9. **Physical and chemical properties** — Physical and chemical properties associated with the substance or mixture. The minimum required information consists of:

o   Appearance (physical state, color, etc.);

o   Flammability (solid, gas);

o   Upper/lower flammability or explosive limits;

o   Flash point;

o   Odor;

o   Odor threshold;

o   Vapor pressure;

o   Vapor density;

o   pH;

o   Relative density;

o   Melting point/freezing point;

o   Solubility(ies);

o   Initial boiling point and boiling range;

o   Evaporation rate;

o   Partition coefficient: n-octanol/water;

o   Auto-ignition temperature;

o   Decomposition temperature; and

o   Viscosity.

The SDS may not contain every item on the above list because information may not be relevant or is not available. When this occurs, a notation to that effect must be made for that chemical property.

Section 10. Stability and reactivity — Describes the reactivity hazards of the chemical and the chemical stability information. The required information is presented in three parts: reactivity, chemical stability, and other:

Reactivity

o   Description of the specific test data for the chemical(s). This data can be for a class or family of the chemical if such data adequately represent the anticipated hazard of the chemical(s), where available.

Chemical stability

o   Indication of whether the chemical is stable or unstable under normal ambient temperature and conditions while in storage and being handled.

o   Description of any stabilizers that may be needed to maintain chemical stability.

o   Indication of any safety issues that may arise should the product change in physical appearance.

Other

o   Indication of the possibility of hazardous reactions, including a statement whether the chemical will react or polymerize, which could release excess pressure or heat, or create other hazardous conditions. Also, a description of the conditions under which hazardous reactions may occur.

○ List of all conditions that should be avoided (e.g., static discharge, shock, vibrations, or environmental conditions that may lead to hazardous conditions).

○ List of all classes of incompatible materials (e.g., classes of chemicals or specific substances) which the chemical could react with to produce a hazardous situation.

○ List of any known or anticipated hazardous decomposition products that could be produced because of use, storage, or heating. (Hazardous combustion products should also be included in Section 5 Fire-Fighting Measures.)

Section 11. Toxicological information — Identifies toxicological and health effects information or indicates that such data are not available. The required information consists of:

○ Information on the likely routes of exposure (inhalation, ingestion, skin and eye contact).

○ An indication if the information is unknown.

○ Description of the delayed, immediate, or chronic effects from short- and long-term exposure.

○ The numerical measures of toxicity (e.g., acute toxicity estimates such as the LD50 (median lethal dose)) - the estimated amount [of a substance] expected to kill 50 percent of test animals in a single dose.

○ Description of the symptoms associated with exposure from the lowest to the most severe exposure.

○ Indication of whether the chemical is listed in the National Toxicology Program (NTP)

Report on Carcinogens (latest edition) or has been found to be a potential carcinogen in the International Agency for Research on Cancer (IARC) Monographs (latest editions) or found to be a potential carcinogen by OSHA

Other information listed in this section includes whether the substance is a carcinogen (known to cause cancer), and any medical conditions that may be aggravated by exposure.

Section 12*. Ecological information — Provides information to evaluate the environmental impact of the chemical(s) if it were released to the environment. The information may include:

o   Data from toxicity tests performed on aquatic and/or terrestrial organisms, where available.

o   Whether there is a potential for the chemical to persist and degrade in the environment either through biodegradation or other processes, such as oxidation or hydrolysis.

o   Results of tests of bioaccumulation potential, making reference to the octanol-water partition coefficient (Kow) and the bioconcentration factor (BCF), where available.

o   The potential for a substance to move from the soil to the groundwater (indicate results from adsorption studies or leaching studies).

o   Other adverse effects (e.g., environmental fate, ozone layer depletion potential, photochemical ozone creation potential,

SAFETY DATA SHEETS

59

endocrine disrupting potential, and/or global warming potential).

Section 13*. Disposal considerations — Provides guidance on proper disposal practices, recycling or reclamation of the chemical(s) or its container, and safe handling practices, including:

○ Description of appropriate disposal containers to use.

○ Recommendations of appropriate disposal methods to employ.

○ Description of the physical and chemical properties that may affect disposal activities.

○ Language discouraging sewage disposal. Any special precautions for landfills or incineration activities.

Section 14*. Transport information — Provides guidance on classification information for shipping and transporting of hazardous chemical(s) by road, air, rail, or sea:

○ UN number (i.e., four-figure identification number of the substance).

○ UN proper shipping name.

○ Transport hazard class(es).

○ Packing group number, if applicable, based on the degree of hazard.

○ Environmental hazards (e.g., identify if it is a marine pollutant according to the International Maritime Dangerous Goods Code (IMDG Code)).

○ Guidance on transport in bulk (according to Annex II of MARPOL 73/783 and the International Code for the Construction and

Equipment of Ships Carrying Dangerous Chemicals in Bulk (International Bulk Chemical Code (IBC Code)).

o   Any special precautions which an employee should be aware of or needs to comply with, in connection with transport or conveyance either within or outside their premises and indicate when information is not available.

Section 15*. Regulatory information — Identifies the safety, health, and environmental regulations specific for the product that is not indicated anywhere else on the SDS, including any national and/or regional regulatory information.

Section 16. Other information — This section indicates when the SDS was prepared or when the last known revision was made. The SDS may also state where the changes have been made to the previous version. Other useful information also may be included here.

*Note: Since other Agencies regulate this information, OSHA will not be enforcing Sections 12 through 15 (29 CFR 1910.1200(g)(2)).

Where to Find the SDS

SDSs must be readily accessible to you in your work area on all shifts.

The location of SDSs should be covered in your employer's training program. Your supervisor can always help you with SDS questions.

If you travel between work sites during a work shift, the SDSs may be kept at a central location at the primary work-place facility as long as you have immediate access to the information.

What if I Cannot Find the SDS for a Chemical?

If you can't find the SDS for a certain chemical, inform your supervisor at once.

Do not use the chemical if you do not have an SDS for reference, or if you do not fully understand the potential hazards and how to protect yourself.

Can SDSs Be Stored on a Computer?

Your employer may keep SDSs in any format as long as the employees have immediate access to the information without leaving their work area and a back-up is available in the case of a power outage or other emergency.

Your employer must train you on how to work the system to get to the SDSs.

Do All Products Need to Have an SDS?

Not all products must have an SDS:

- Non-hazardous chemicals do not need an SDS.

- Household consumer products which are used in the workplace in the same manner that a consumer would use them do not need an SDS.

In order to fall under the consumer product exemption the duration and frequency of use would have to be the same as that of a normal consumer using that product.

Do SDSs Expire?

An SDS does not have a "shelf life." OSHA says that a new SDS is to be developed when new information about the chemical is discovered. If there is no new information, then the SDS is still valid.

In Canada, there is a workplace chemical safety rule called the Workplace Hazardous Materials Information System (WHMIS). WHMIS requires that the supplier or employer update or replace the SDS every three years. This is for Canada only, but often in the U.S. employers assume that they must also replace their SDSs every three years.

OSHA requires that when chemical manufacturers or import-ers become newly aware of any significant information regarding the hazards of the chemical or ways to protect against the hazards, the SDS must be updated within three months or before the chemical is introduced into the work-place again.

Can My Doctor See the SDSs?

You may request copies of SDSs to share with your doctor. Your doctor may wish to know more about the chemical hazards that you are being exposed to, or the signs and symptoms of exposure.

Employee _____

Instructor _____

Date _____

Location _____

Safety Data Sheets Self-Check

1. The SDS is your primary source for information about the chemical you are working with.

    a. TRUE

    b. FALSE

2. On the SDS, Section 4. First-aid measures includes:

    a. Information on first-aid instructions by relevant routes of exposure.

    b. A description of the most important symptom or effect.

    c. Recommendations for immediate care.

    d. All of the above.

3. Section _____ Exposure controls/personal protection covers exposure limits and personal protective measures.

4. You must be informed of where to find the SDS for your work area during your initial HazCom training.

    a. TRUE

    b. FALSE

SAFETY DATA SHEETS

65

5. Which of the following statements is true?

    a. If an SDS is missing, you may continue to work with the chemical until one is obtained.

    b. If you have questions about an SDS, you should ask your supervisor.

    c. Employers may not store SDSs electronically.

    d. Your employer must supply SDSs in languages other than English.

6. All chemical products in the workplace must have an SDS.

    a. TRUE

    b. FALSE

7. OSHA requires that an SDS be reviewed and updated every three years.

    a. TRUE

    b. FALSE

Employee Information & Training

Each employee who may be "exposed" to hazardous chemicals must be provided information and training prior to his or her initial assignment to work with a hazardous chemical, and whenever the hazards change.

Under the rule, "exposure" or "exposed" means that an employee is subjected to a hazardous chemical in the course of employment through any route of entry (inhalation, ingestion, skin contact, or absorption) and includes potential (e.g., accidental or possible) exposure.

But you must receive training on all chemical hazards that you may be exposed to during your course of employment, regardless of the route of exposure or entry.

What Is Effective Training?

OSHA requires that "effective" training must be provided:

- At the beginning of each new assignment involving hazardous chemicals, and

- Whenever a new physical or health hazard is introduced.

By "effective," OSHA means that the information and training program must provide employees with the knowledge they need, and that the employees carry the knowledge from the training into their daily jobs.

What Must I Know?

OSHA expects your employer to provide you with the information and training you need to stay safe on the job. Once you are armed with that knowledge, you can take precautions to keep yourself and co-workers safe.

Your employer is required to inform you of the following:

- Requirements of the standard;
- Places where hazardous chemicals are present in your work area;
- The location and availability of the:
  - Written program,
  - Chemical inventory, and
  - Safety data sheets (SDS);
- How to access the SDSs for your work area;
- How to read an SDS;
- How to read the GHS-style container labels;

- The type of in-house labeling system used, if different; and

- Specific hazardous chemicals in the employees' work areas.

OSHA also requires that the training be understood by the employee. This may mean that your employer must provide training in a language other than English, or through an interpreter.

Your employer must also give you the opportunity to ask questions about your training and about the hazards of the chemicals you work with.

Chemical Specific Training

OSHA allows employers to train on each individual hazardous chemical or to train on a general chemical hazard "group," such as corrosives or flammables.

So if your employer has a lot of one type of hazardous chemical, you may be trained on the hazards of that group of chemicals, rather than receiving training on each individual chemical.

Training by hazard group is usually the method preferred by companies whose employees work with so many chemicals that it is nearly impossible to train them in each specific one.

Regardless if your employer decides to train by individual chemical or by hazard classes, your hazardous chemical training must include:

- How to detect the presence or release of a hazardous chemical.

- The physical and health hazards of those chemicals.

- Measures you can use to protect yourself against these hazards, including:

  o The use of safe work practices,

  o Emergency procedures, and

   o   Proper personal protective equipment required.

When Must Re-Training Be Done?

OSHA does not specify that employers must perform refresher training, or when refresher training must be given.

However, OSHA does require that additional training is to be done whenever a new physical or health hazard is introduced into the work area. If a new solvent is brought into the workplace, and you have already had training on similar chemicals, no new training is required.

If a new process is introduced, with different chemical hazards than the ones you have been trained on, then new training must be provided for employees working in those areas.

General information, such as the basics of the HCS, would stay with an employee. In this case, the employee would not have to receive additional training, unless the hazards of the new position called for it.

It is always the employer's responsibility to make sure that employees are properly trained and equipped with the knowledge and information necessary to do their jobs safely.

Does the Standard Apply to Office Environments?

Office workers who encounter hazardous chemicals only in non-routine, isolated instances do not require training.

OSHA considers most office products (such as pens, pencils, and adhesive tape) to be exempt under the provisions of the rule, either as articles or as consumer products. OSHA has previously stated that intermittent or occasional use of a copying machine does not result in coverage under the rule. However, if an employee handles the chemicals to service the machine, or operates it for long periods of time, then the program would have to be applied.

Employee _____

Instructor _____

Date _____

Location _____

Employee Information & Training Self-Check

1. Each employee who may be exposed to hazardous chemicals must be provided information and training.

   a. TRUE

   b. FALSE

2. OSHA requires that employers provide "effective" training:

   a. At the beginning of each new assignment involving hazardous chemical.

   b. In a language the employee receives work instructions in.

   c. Whenever a new physical or health hazard is first introduced.

   d. All of the above.

3. _____ training means that it must provide employers with the knowledge they need, and that the employees apply that knowledge in their jobs.

4. Under HazCom, your employer must inform you of:

   a. The requirements of the standard.

   b. How to read a GHS-style SDS.

   c. How to read a GHS-style container label.

   d. All of the above.

71

5. You must be given the opportunity to ask questions during HazCom training.

   a. TRUE

   b. FALSE

6. Your employer must provide you with training when all of the following occur EXCEPT:

   a. Each time the chemical inventory is updated.

   b. At the time of your initial assignment to a job.

   c. Whenever a new hazard is introduced to your work area.

   d. Whenever exposure controls, such as work practices or emergency procedures, change.

7. Your employer must always train you on the hazards of each individual chemical, and not by hazard groups.

   a. TRUE

   b. FALSE

8. According to OSHA, HazCom refresher training must occur every year.

   a. TRUE

   b. FALSE

Working With Chemical Hazards

If you work with hazardous chemicals, be aware that every hazardous chemical substance handled, whether it is a liquid, solid, vapor, or dust, could cause you harm if you do not understand the hazards and are not properly protected.

Before working with hazardous chemicals at your job:

- Get HazCom training from your employer.

- Review the company's written program.

- Learn about the hazards of the chemicals you work with.

- Find and review the SDSs for the chemicals you will be using.

- Read the container label each time you use that chemical.

What Is Exposure?

If you come into contact with a chemical that presents a physical or health hazard in the course of your employment, you are said to be "exposed." This includes the potential for exposure - meaning accidental or possible exposure.

There are two types of chemical exposures:

- **Acutely toxic** chemicals can cause harm after a single exposure.

- **Chronically toxic** chemicals can cause harm after repeated exposures.

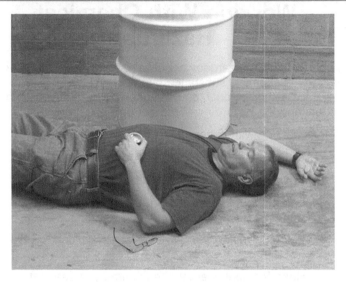

How Can Chemicals Hurt Me?

Chemicals vary greatly in their toxicity, depending upon the amount, frequency, and duration of exposure. Identifying the chemical and reading about the hazards on the SDS and on the container label are critical to your understanding of the risks of exposure.

There are three common routes of entry into the human body:

- Skin contact
- Ingestion
- Inhalation

Skin Contact

Liquids may get on the skin accidentally, through a spill, or through a job process. Some chemicals will cause an external reaction on the skin, such as a burn or an irritation. Other

74

chemicals may pass through the skin and enter the body, possibly with severe results.

Ingestion

Exposure can also occur through swallowing. A chemical that cannot penetrate the skin may enter the body through the mouth. Drinking and eating, smoking, or applying makeup in high risk areas increases the chance of a toxic material accidentally getting into your mouth.

Inhalation

Breathing in toxic dust or vapors is the most difficult exposure method to control. That is because you may inhale a harmful substance without being aware of it. It is hard to take precautions if you do not realize a situation is occurring.

How Does Your Body React?

There are two types of chemical reactions; local reactions or systemic reactions. Local reactions take place at the site of the exposure. For instance, there is usually an immediate reaction on the skin or in the eyes, or in the lungs if inhaled.

Systemic reactions typically occur sometime after exposure, and may involve only certain bodily organs, known as target organs. With systemic reactions, you may not know you are being harmed until it is too late.

Know Your Role in Chemical Spills

Your employer will have a plan for what to do in the event of a chemical spill. Do you understand what steps you are to take if there is an accidental release of hazardous substances?

At a minimum, you should understand how to report a spill, whether you are to try to clean up smaller spills or not, and where emergency contact numbers can be found to report a spill.

Tips for Working With Chemicals

Here are some common-sense tips to follow when working with hazardous chemicals;

- Know the location of the nearest emergency eyewash/ shower station.

- Since certain metals can react with chemicals, always remove jewelry to avoid skin reactions.

- Use face or eye protection if there is a possibility of a hazardous chemical splashing in your face.

- If you wear contact lenses, check with your supervisor about prescription safety goggles.

- Make sure PPE is clean and properly functioning prior to use.

- Understand the limits of the PPE.

- Wash hands, arms, and face thoroughly after working with hazardous chemicals.

Working With Chemical Hazards

- Always wash hands before eating, smoking, or applying makeup.

- Do not eat, drink, smoke, or use personal items in an area where hazardous chemicals are present. It is possible for toxic fumes or particles to enter your body as you swallow or inhale.

- Know the company policy for handling a hazardous chemical spill or leak, and know how to report such a spill or leak.

Notes

Employee _____

Instructor _____

Date _____

Location _____

Working With Chemical Hazards Self-Check

1. Every chemical substance you handle, whether it is a liquid, solid, vapor, or dust, could cause you harm if you aren't properly protected.

   a. TRUE

   b. FALSE

2. Before working with hazardous chemicals at your job you should:

   a. Get HazCom training from your employer.

   b. Review the company written program.

   c. Learn about the hazards of the chemicals you work with.

   d. All of the above.

3. When you come into contact with a chemical that is a physical or health hazard in the course of your employment, you are said to be "exposed."

   a. TRUE

   b. FALSE

4. Acutely toxic chemicals can only cause harm after repeated exposures.

   a. TRUE

   b. FALSE

5. There are three common routes of entry into the human body:

   a. Skin contact

   b. Inhalation

   c. Ingestion

   d. All of the above.

HazCom Terms & Abbreviations

Absorbed dose The amount of a substance that actually enters into the body, usually expressed as milligrams of substance per kilogram of body weight (mg/kg).

Action level The exposure level at which OSHA regulations take effect. Exposure at or above action level is termed occupational exposure.

Acute dose The amount of a substance administered or received over a very short period of time (minutes or hours), usually within 24 hours.

Acute exposure An exposure to a toxic substance which occurs in a short or single time period.

Acute toxicity Acute effects resulting from a single dose of, or exposure to, a substance. Ordinarily used to denote effects in experimental animals.

ALARA As Low As Reasonably Achievable. Often used in reference to reducing exposure to levels that are as low as can be achieved.

Allergic reaction An abnormal immunologic response in a person who has become hypersensitive to a specific substance. Some forms of dermatitis and asthma may be cause by allergic reactions to chemicals.

Allergy An abnormal physical response to chemical and physical stimuli which occurs in about 10 percent of the population.

Alopecia The loss of hair, which can be caused by chemical exposure.

Anesthetic effect The temporary loss of feeling induced by certain chemical agents, which reduce the ability to feel pain or other sensations.

Anhydrous Not containing any water molecules. Some anhydrous chemicals are very water reactive.

Anorexia Loss of appetite. Can be a symptom of exposure to a given chemical.

Anosmia Loss of smell. Can be a symptom of exposure to a given chemical.

Antidote A remedy that counters the effects of a toxin or poison.

AQTX Aquatic toxicity, the adverse effects on fresh or salt water life forms that result from exposure to a toxic substance.

Asphyxiant A vapor or gas that can cause unconsciousness or death by suffocation.

Aspiration hazard The danger of drawing a fluid into the lungs and causing an inflammatory response.

Asymptomatic Not causing or showing signs of exposure or symptoms of disease.

Autoignition temperature Temperature at which a liquid will self-ignite and sustain combustion without outside spark or flame.

BCF Bioconcentration factor.

BEI Biological Exposure Index a maximum level set by the ACGIH of substance in the blood, exhaled air or urine.

Benign Referring to a non-malignant tumor which does not metastasize or invade tissue. Benign tumors may still be lethal due to pressure on organs.

BOD/COD Biochemical oxygen demand/chemical oxygen demand.

Bradycardia Reduces heart rate of 60 beats per minute or less, resulting in shortness of breath, dizziness, fatigue, or fainting.

Bronchitis Acute or chronic inflammation of the bronchial tubes, resulting in a fluid-producing cough.

Carcinogen A chemical substance or a mixture of chemical substances which induce cancer or increase its incidence

Carcinoma A malignant tumor found in the membranes that line and protect the organs which can spread to other tissues or organs.

CAS Chemical Abstract Service.

Cataract A clouding of the lens of the eye or its surrounding transparent membrane that obstructs the passage of light.

Ceiling limit Maximum amount of a toxic substance allowed in workroom air any time during the day.

Chemical asphyxiant Substances that prevent the body from receiving or using an adequate oxygen supply. Carbon monoxide and cyanide are examples.

Chemical family A group of compounds with related chemical and physical properties. For example, acetone, methyl ethyl ketone (MEK), and methyl isobutyl ketone (MIBK) are three members of the"ketone" family.

Chemical identity A name that will uniquely identify a chemical. This can be a name that is in accordance with the nomenclature systems of the International Union of Pure and Applied Chemistry (IUPAC) or the Chemical Abstracts Service (CAS), or a technical name.

Chemical name The scientific designation of a chemical in accordance with the nomenclature system developed by the International Union of Pure and Applied Chemistry (IUPAC) or the Chemical Abstracts Service (CAS) rules of nomenclature or a name that will clearly identify the chemical for the purpose of conducting a hazard classification.

Chemical pneumonitis Inflammation of the lungs caused by accumulation of fluids due to chemical irritation.

Chemical reactivity A chemical's tendency to react with other materials.

Chronic effect An adverse effect on a human or animal body, with symptoms that develop slowly over a long period of time or that recur frequently.

HAZCOM TERMS & ABBREVIATIONS

Chronic toxicity Adverse (chronic) effects resulting from repeated doses of or exposures to a substance over a relatively prolonged period of time. Ordinarily used to denote effects in experimental animals.

Classification To identify the hazards of a chemical and decide whether the chemical will be classified as hazardous by comparing the data with the criteria for health and physical hazards.

CNS Central nervous system, composed of the brain and spinal cord.

CNS depression Lowered sensitivity level or loss of sensation in the central nervous system, usually due to exposure to a particular chemical hazard or anesthetic.

Combustible liquid A combustible liquid according to OSHA is any liquid having a flashpoint at or above 100°F (37.8°C) Combustible liquids shall be divided into two classes as follows: (i) Class II liquids shall include those with flashpoints at or above 100°F (37.8°C) and below 140°F (60°C), except any mixture having components with flashpoints of 200°F. (93.3°C) or higher, the volume of which make up 99 percent or more of the total volume of the mixture. (ii) Class III liquids shall include those with flashpoints at or above 140°F (60°C) Class III liquids are subdivided into two subclasses: (a) Class IIIA liquids shall include those with flashpoints at or above 140°F (60°C) and below 200°F (93.3°C), except any mixture having components with flashpoints of 200°F (93.3°C), or higher, the total volume of which make up 99 percent or more of the total volume of the mixture. (b) Class IIIB liquids shall include those with flashpoints at or above 200°F (93.3°C). (iii) When a combustible liquid is heated for use to within 30°F (16.7°C) of its flashpoint, it shall be handled in accordance with the requirements for the next lower class of liquids.

Compressed gas A gas which when packaged under pressure is entirely gaseous at -50°C; including all gases with a critical temperature ≤ -50°C.

Conjunctivitis Inflammation of the conjunctiva, the delicate membrane that lines the eyelids and covers the eyeballs.

Contact sensitizer A substance that will induce an allergic response following skin contact. Also 'skin sensitizer.'

Corneal/conjunctival burns Burns to the transparent membrane covering the eyeball and lining the eyelids.

Corrosive A chemical that causes visible destruction of, or irreversible alterations in, living tissue by chemical action at the site of contact.

Cryogenics Materials which exist at extremely low temperatures, such as liquid nitrogen.

Cutaneous Pertaining to or affecting the skin.

CVS Cardio-vascular system.

Cyanosis The bluish or purplish discoloration of the skin due to the lack of oxygen in the blood.

Cytotoxin A substance having a toxic effect on cells, or on the cells of a specific organ of the body.

Dermal By or through the skin.

Dermal sensitization An exposure of an agent to skin which results in an immune response. Subsequent exposure will often induce a much stronger (secondary) immune response.

Dermal toxicity Adverse toxic effects resulting from skin exposure to a substance.

Dermatitis Inflammation of the skin from any cause.

Dermatosis All cutaneous abnormalities more serious than inflammation, such as folliculitis, acne, pigmentary changes, nodules, or tumors.

Dissolved gas A gas which when packaged under pressure is dissolved in a liquid phase solvent. (A GHS reference.)

Dose The amount of a substance received at one time. Dose is usually expressed as administered or absorbed dose (e.g., milligrams material/kilogram of body weight).

Dose-response relationship The correlation between the amount of exposure to an agent or chemical and the result and effects upon a person.

Dysosmia The distortion or lack of the sense of smell.

Dyspnea A sense of difficulty in breathing; shortness of breath.

EC_{50} The effective concentration of a substance that causes 50% of the maximum response.

EC number or (ECN) A reference number used by the European Communities to identify dangerous substances, in particular those registered under EINECS.

Edema An abnormal infiltration and accumulation of fluids in connective tissues or body cavities.

EINECS The "European Inventory of Existing Commercial Chemical Substances".

Embolism A sudden partial or total blockage of a blood vessel caused by a blood clot, air, bacteria, foreign material, or other bodily substance.

Emergency release An emergency spill or release refers to the after effects of an unintended release of hazardous, toxic, or explosive substances.

End point A GHS term referring to physical, health and environmental hazards of a hazard class and category.

Epidemiology The branch of science concerned with the study of human disease in specific populations, in order to develop information about the cause of disease and identify preventive measures.

ErC50 The same as EC50 in terms of reduction of growth rate.

Evaporation The process by which a liquid is changed into a vaporous state.

Explosive A chemical that causes a sudden, almost instantaneous release of pressure, gas, and heat when subjected to sudden shock, pressure, or high temperature.

Explosive article An article containing one or more explosive substances. (A GHS reference.)

Explosive limit The amounts of vapor in air which form explosive mixtures. Explosive limits are expressed as Lower Explosive Limits and Upper Explosive Limits; these give the range of vapor concentrations in air which will explode if heat is added. Explosive limits are expressed as percent of vapor in air.

Explosive substance A solid or liquid substance or mixture of substances which is in itself capable by chemical reaction of producing gas at such a temperature and pressure and at such a speed as to cause damage to the surroundings. Pyrotechnic substances are included even when they do not involve gases. (A GHS reference.)

Exposure or Exposed When an employee is subjected in the course of employment to a chemical that is a physical or health hazard, and includes potential (e.g. accidental or possible) exposure. "Subjected" in terms of health hazards includes any route of entry (e.g. inhalation, ingestion, skin contact or absorption.)

Exposure limit A chemical's safe concentration in workplace air. At this level or lower most workers can be exposed without harmful effects.

Extremely hazardous substance Any one of over 300 hazardous chemicals on a list compiled by ERA to provide a focus for State and local emergency planning activities.

Eye irritation Changes in the eye following the application of test substance to the front surface of the eye, which are fully reversible within 21 days of application.

F Fahrenheit is a scale for measuring temperature. On the Fahrenheit scale, water boils at 212° F and freezes at 32° F.

f/cc Fibers per cubic centimeter of air.

Fibrosis The formation of fibrous connective or scar tissue due to disease or injury.

Fire point The lowest temperature at which a material can evolve vapors fast enough to support continuous combustion.

Flammable aerosol An aerosol that yields a flame projection for more than 18" at full valve opening, or a flash back (a flame extending back to the valve) at any degree of valve opening.

Flammable gas A gas having a flammable range with air at 20ºC and a standard pressure of 101.3 kPa.

Flammable limits The concentration of vapors in the air below which or above which ignition cannot occur. See Explosive Limits.

Flammable liquid A liquid having a flash point of not more than 93°C.

Flammable solid A solid which is readily combustible, or may cause or contribute to fire through friction.

Flash Point (FP) The lowest temperature at which the vapor of a substance will catch on fire, even momentarily, if heat is applied. Provides an indication of how flammable a substance is.

Foreseeable emergency Any potential occurrence such as, but not limited to, equipment failure, rupture of containers, or failure of control equipment which could result in an uncontrolled release of a hazardous chemical into the workplace.

Fumes Fumes are formed by processing, such as combustion, sublimation, or condensation. The term is generally applied to the metal oxides of such metals as zinc, magnesium, or lead.

g Gram is a metric unit of weight. One ounce U.S. (avoirdupos) is about 28.4 grams.

Gas A substance which (i) at 50 °C has a vapor pressure greater than 300 kPa; or (ii) is completely gaseous at 20 °C at a standard pressure of 101.3 kPa.

Gastroenteritis Inflammation of the membrane lining the stomach and intestines, leading to nausea, abdominal pain, and diarrhea.

GHS The Globally Harmonized System of Classification and Labelling of Chemicals.

g/kg Grams per kilogram is an expression of dose used in oral and dermal toxicology testing to denote grams of a substance dosed per kilogram of animal body weight. Also see kg (kilogram).

Hazard category the division of criteria within each hazard class, *e.g.*, oral acute toxicity and flammable liquids include four hazard categories. These categories compare hazard severity within a hazard class and should not be taken as a comparison of hazard categories more generally.

Hazard class The nature of the physical or health hazards, *e.g.*, flammable solid, carcinogen, oral acute toxicity.

Hazard warning Any words, pictures, symbols, or combination thereof appearing on a label or other appropriate form of warning that conveys the hazard(s) of the chemical(s) in the container(s).

Hazard Not Otherwise Classified (HNOC) Means an adverse physical or health effect identified through evaluation of scientific evidence during the classification process that does not meet the specified criteria for the physical and health hazard classes addressed in the HCS.

Hazard statement A statement assigned to a hazard class and category that describes the nature of the hazard(s) of a chemical, including, where appropriate, the degree of hazard.

Hazardous atmosphere An atmosphere that may expose employees to the risk of death, incapacitation, impairment of ability to self-rescue (escape unaided), injury, or acute illness.

Hazardous chemical Any chemical which is classified as a physical hazard or a health hazard, or an unclassified hazard as defined in the HCS.

Hazardous ingredient A category of ingredients which, under the Hazardous Products Act, must be listed in the Hazardous Ingredients section of an SDS if: it meets the criteria for a controlled product; it is on the Ingredient Disclosure List; there is no toxicological information available; or the supplier has reason to believe it might be hazardous.

HBV Hepatitis B Virus.

Health hazard A chemical that is classified as posing one of the following hazardous effects: acute toxicity; skin corrosion or irritation; serious eye damage or eye irritation; respiratory or skin sensitization; germ cell mutagenicity; carcinogenicity; reproductive toxicity; specific target organ toxicity (single or repeated exposure); or aspiration hazard.

Hematology The study of blood.

Hematoma A blood clot under the surface of the skin.

Hematopoietic system The blood forming mechanism of the human body.

Hepatotoxin A substance which can have a harmful effect on the liver.

Highly toxic The greatest level of toxicity that a chemical can have as defined by OSHA in Appendix A of the Hazard Communication Standard, 29 CFR 1910.1200.

Hyperplasia Increase in volume of a tissue or organ caused by the growth of new cells.

Hypoxia Deficiency of oxygen reaching tissues through the blood, caused by a lack of or displacement of oxygen being inhaled.

IDLH Immediately Dangerous to LIfe or Health. As defined by NIOSH, this represents a hazardous atmosphere from which one could escape within 30 minutes without any escape-impairing symptoms or any irreversible health effects. The NIOSH definition addresses airborne concentration only.

Ignitable A solid, liquid or compressed gas which is capable of being set afire.

Ignition source Anything which provides flame, heat, or spark sufficient to cause combustion or an explosion.

Ignition temperature The lowest temperature at which a substance will catch on fire and continue to burn. The lower the ignition temperature, the more likely the substance is going to be a fire hazard.

Immediate use Means that the hazardous chemical will be under the control of and used only by the person who transfers it from a labeled container and only within the work shift in which it is transferred.

Impervious A material that does not allow another substance to pass through or penetrate it.

In Vitro Outside a living organism (e.g., in a test tube).

Incidental release An "incidental release" is a release of a hazardous substance which does not pose a significant safety or health hazards to employees in the immediate vicinity or to the worker cleaning it up, nor does it have the potential to become an emergency.

Incompatible Materials which could cause dangerous reactions from direct contact with one another.

Inflammable Same as Flammable.

Inhibitor A chemical added to another substance to prevent an unwanted chemical change.

Initial boiling point The temperature of a liquid at which its vapor pressure is equal to the standard pressure (101.3 kPa), i.e., the first gas bubble appears.

Insol or Insoluble Incapable of being dissolved in liquid.

Irritant A chemical which is not a corrosive, that causes a reversible inflammatory effect on living tissue by chemical reaction at the site of contact.

Jaundice A yellowish pigmentation of the skin, tissues, and body fluids usually caused by an improperly functioning liver.

Ketosis An abnormal increase of ketone in the body, which can lead to a lowered blood pH level eventually causing coma and even death.

Kilogram (kg) 1000 grams. One kilogram equals about 2.2 pounds.

L Liter is a metric unit of capacity. A U.S. quart is about 9/10 of a liter.

LC_{50} The concentration of a substance in air that causes death in 50% of the animals exposed by inhalation. A measure of acute toxicity.

LD_{50} The dose that causes death in 50% of the animals exposed by swallowing a substance. A measure of acute toxicity.

$L(E)C_{50}$ LC50 or EC50.

Lesion An abnormal change in the structure of tissue or organs due to disease or injury, usually characterized by a "break" in the tissue.

Lethal concentration-50% (LC50) A concentration of a substance in the air that will kill 50% of the test animals which inhale it.

Lethal concentration low Lethal concentration low is the lowest concentration of a gas or vapor capable of killing a specified species over a specified time period.

Lethal dose (LD) Lethal dose is the quantity of a substance being tested that will kill.

Lethal dose-50% (LD50) The dose of a substance, given orally, applied to the skin, or injected, which will kill 50% of the test animals receiving it. A rough measure of acute toxicity.

Lethal dose low (LDL) Lethal dose low is the lowest administered dose of a material capable of killing a specified test species.

Liquiefied gas A gas which when packaged under pressure, is partially liquid at temperatures above –50°C. A distinction is made between. (i) High pressure liquefied gas: a gas with a critical temperature between -50°C and +65°C; and (ii) Low pressure liquefied gas: a gas with a critical temperature above +65°C.

LOAEL The Lowest Observed Adverse Effect Level, i.e., the lowest dose which produces an observable adverse effect.

Lower explosive limit (LEL) Refers to the lowest concentration of gas or vapor by percent volume in air that explodes if an ignition source is present at ambient temperatures.

Lower flammable limit (LFL) Refers to the lowest concentration of gas or vapor by percent volume in air that burns if an ignition source is present at ambient temperatures.

Malignant Tending to infiltrate, metastasize, and cause fatality; i.e. cancer that spreads to other areas of the body.

Material causing immediate and serious toxic effects Classified under "Poisonous and Infectious Material" as toxic or very toxic based on information such as the LD_{50} or LC_{50}.

Material causing other toxic effects Classified under "Poisonous and Infectious Material" as a material causing toxic effects such as skin or respiratory sensitization, carcinogenicity, mutagenicity, etc.

Malignant As applied to a tumor, meaning cancerous and capable of invading surrounding tissue.

Melting point The temperature at which a solid substance changes to a liquid state.

Metabolism The conversion of a chemical from one form to another within the body; also referred to as biotransformation.

Metabolite A chemical produced during metabolism.

Metastasis Secondary growth of a malignant tumor; spread of a disease from the original site to other parts of the body.

mg/kg A way of expressing dose: milligrams (mg) of a substance per kilogram (kg) of body weight. Example: a 100 kg person given 10,000 mg of a substance would be getting a dose of 100 mg/kg (10,000 mg/100kg).

mg/m³ A way of expressing the concentration of a substance in air: milligrams (mg) of substance per cubic meter (m^3) of air.

Milligram One one-thousandth of a gram.

Mld Mild

Miscible Capable of being mixed in any ratio without separation of the components.

Mixture A combination or a solution composed of two or more substances in which they do not react.

ml Milliliter; a metric unit of volume. There are 1,000 milliliters in one liter. 1 teaspoon = 5 milliliters.

Molecular weight The sum of the total of the molecular weights of all of the atoms which make up a molecule.

Mucous membrane Mucous-secreting membrane lining the hollow organs of the body, for example, the nose, mouth, stomach, intestines, bronchial tubes, and urinary tract.

Mutagen An agent giving rise to an increased occurrence of mutations in populations of cells and/or organisms. (A GHS reference.)

Mutagenic Capable of changing cells in a way that future cell generations are affected. Mutagenic substances are usually considered suspect carcinogens.

Mutation A permanent change in the amount or structure of the genetic material in a cell.

MW See molecular weight.

NA number Numbers assigned to potentially hazardous materials or class of materials assigned by Transport Canada and the US Department of Transportation, to which a UN number has not yet been assigned.

Narcosis Unconsciousness or stupor caused by the influence of narcotics or other chemicals.

Nausea Tendency to vomit, feeling of sickness at the stomach.

Neonatal The first 4 weeks after birth.

Nephrotoxin A substance that causes injury to the kidneys.

Neurotoxin A material that affects the nerve cells and may produce emotional or behavioral abnormalities.

NOAEL No Observable Adverse Effect Level.

NOEC The "no observed effect concentration."

Noncombustible Material that will not burn or support combustion, or release flammable vapors when subjected to heat or fire.

Non-Flammable Not easily ignited, or if ignited, not burning rapidly.

NPIRS National Pesticide Information Retrieval System is an automated data base operated by Purdue University containing information on EPA registered pesticides, including reference file material safety data sheets.

NTP National Toxicology Program, coordinates toxicology research and testing activities within the U.S. Department of Health and Human Services. The NTP publishes a Report on Carcinogens.

Nuisance dust Dusts or airborne particles which have no or little history of adverse effect on the lungs or which do not produce toxic effects or organic diseases when exposures are kept to a reasonable level.

Occupational exposure limits Maximum allowable concentrations of toxic substances in workroom air to protect workers who are exposed to toxic substances over a working lifetime.

Odor threshold The lowest concentration of a substance's vapor, in air, that can be smelled. Odor thresholds are highly variable depending on the individual who breathes the substance and the nature of the substance.

Oral LD 50 Oral Lethal Dose 50; the concentration of a substance administered by mouth that will produce death in 50 percent of the animals tested.

HAZCOM TERMS &
ABBREVIATIONS

Oral toxicity Adverse effects that result from taking a substance into the body via the mouth.

Organic peroxide A liquid or solid organic substance which contains the bivalent -0-0- structure and may be considered a derivative of hydrogen peroxide, where one or both of the hydrogen atoms have been replaced by organic radicals. The term also includes organic peroxide formulation (mixtures).

ORM-A A DOT hazard classification applied to a material which has an anesthetic, irritating, noxious, toxic, or other similar property and which can cause extreme annoyance or discomfort to passengers and crew in the event of leakage during transportation.

ORM-B A DOT hazard classification applied to a material (including a solid when wet with water)capable of causing significant damage to a transport vehicle or vessel by leaking during transportation.

ORM-C A DOT hazard classification applied to a material that has other inherent characteristics not described as an ORM-A or ORM-B, but that make it unsuitable for shipment unless properly identified and prepared for transportation.

ORM-D A DOT hazard classification applied to a material such as a consumer commodity which, though otherwise subject to the regulations of the DOT hazard classification system, presents a limited hazard during transportation due to its form, quantity, and packaging.

ORM-E A DOT hazard classification applied to a material which is not included in any other hazard class but which is subject to the requirements of the DOT regulations. Materials in this class include "Hazardous Waste" and other hazardous materials.

Oxidation In a literal sense, oxidation is a reaction in which a substance combines with oxygen provided by an oxidizer or oxidizing agent.

Oxidizer A material which may cause the ignition of combustible materials without the aid of an external source of ignition or which, when mixed with combustible materials, increases the rate of burning of these materials when the mixtures are ignited.

Oxidizing agent A chemical or substance that brings about an oxidation reaction. The agent may (1) provide the oxygen to the substance being oxidized (in which case the agent has to be oxygen or contain oxygen), or (2) it may receive electrons being transferred from the substance undergoing oxidation (chlorine is a good oxidizing agent for electron-transfer purposes, even though it contains no oxygen).

Oxidizing gas Any gas which may, generally by providing oxygen, cause or contribute to the combustion of other material more than air does.

Oxidizing liquid A liquid which, while in itself not necessarily combustible, may, generally by yielding oxygen, cause, or contribute to, the combustion of other material.

Oxidizing solid A solid which, while in itself not necessarily combustible, may, generally by yielding oxygen, cause, or contribute to, the combustion of other material.

Oxygen-deficient atmosphere An atmosphere having an oxygen concentration of less than 19.5% by volume.

Particulate matter Commonly known as aerosol, particulate matter is the suspension of fine solid or liquid particles in the air, such as a dust, fog, fume, mist, smoke, or spray.

PEL Permissible Exposure Limit. This is the OSHA-mandated exposure limit.

Percent volatile Percent volatile by volume is the percentage of a liquid or a solid (by volume) that will evaporate at an ambient temperature of 70°F (unless some other temperature is specified). Examples: butane, gasoline, and paint thinner (mineral spirits) are 100 percent volatile; their individual evaporation rates vary, but, in time, each will evaporate completely.

Percutaneous Effected or performed through the skin, such as a substance that can pass through the skin.

HAZCOM TERMS & ABBREVIATIONS

Permeability The ability to pass or penetrate a substance or membrane.

pH A measure of how acid or how caustic (basic) a substance is on a scale of 1-14. pH 1 indicates that a substance is very acid; pH 7 indicates that a substance is neutral; and pH 14 indicates that a substance is very caustic (basic).

Photosensitization (contact) After exposure to some chemical substance(s), the skin, upon exposure to light, may swell or exhibit dermatitis.

Physical hazard A chemical that is classified as posing one of the following hazardous effects: explosive; flammable (gases, aerosols, liquids, or solids); oxidizer (liquid, solid, or gas); self-reactive; pyrophoric (liquid or solid); self-heating; organic peroxide; corrosive to metal; gas under pressure; or in contact with water emits flammable gas.

Pictogram A composition that may include a symbol plus other graphic elements, such as a border, background pattern, or color, that is intended to convey specific information about the hazards of a chemical. Eight pictograms are designated under this standard for application to a hazard category.

PLHCP Physician or Other Licensed Health Care Provider.

PMCC Pensky-Martens Closed Cup. See flashpoint.

Pneumoconiosis A condition of the lung in which there is permanent deposition of particulate matter and the tissue reaction to its presence. It may range from relatively harmless forms of iron oxide deposition to destructive forms of silicosis.

Pneumonitis Inflammation of the lungs, which may be caused by inhalation of chemical irritants.

Poison, Class A A DOT term for extremely dangerous poison, poisonous gases or liquids that, in very small amounts, either as gas or as vapor of the liquid, mixed with air, are dangerous to life. Examples: phosgene, cyanogen, hydrocyanic acid, nitrogen peroxide.

Poison, Class B A DOT term for liquid, solid, paste or semi-solid substances, other than Class A poisons or irritating materials, that are known (or presumed on the basis of animal tests) to be so toxic to humans that they are a hazard to health during transportation.

PPB Parts per billion.

ppm Parts per million. Generally used to express small concentrations of one substance in a mixture.

Precautionary statement A phrase that describes recommended measures that should be taken to minimize or prevent adverse effects resulting from exposure to a hazardous chemical, or improper storage or handling.

Product identifier A unique name or number used for a hazardous chemical on a label or in the SDS which permits cross-references to be made among the list of hazardous chemicals, the label, and the SDS.

Pul or Pulmonary Related to, or associated with, the lungs.

Pulmonary edema The abnormal accumulation of fluid in the tissues and air spaces of the lungs.

Pyrophoric A chemical than will ignite spontaneously in air at temperature of 130 ° F (54.4 ° C) or below.

Pyrophoric gas A chemical in a gaseous state that will ignite spontaneously in air at a temperature of 130 degrees F (54.4 degrees C) or below.

Pyrophoric liquid A liquid which, even in small quantities, is liable to ignite within five minutes after coming into contact with air.

Pyrophoric solid A solid which, even in small quantities, is liable to ignite within five minutes after coming into contact with air.

Pyrotechnic substance A substance or mixture of substances designed to produce an effect by heat, light, sound, gas or smoke or a combination of these as the result of non-detonative, self- sustaining exothermic (heat-related) chemical reactions.

QRA Quantitative Risk Assessment. Risk control procedures that can be quantified.

QSAR Quantitative structure-activity relationships.

R-Phrases Standardized risk phrases that are required on labels and safety data sheets for hazardous chemicals in the European Union. They appear as a letter followed by one or more numbers.

Reactive See Unstable.

Reactivity The ability of a substance to undergo change, usually by combining with another substance or by breaking down. Certain conditions, such as heat and light, may cause a substance to become more reactive. Highly reactive substances may explode.

Readily combustible solid Powdered, granular, or pasty substance or mixture which is dangerous if it can be easily ignited by brief contact with an ignition source, such as a burning match, and if the flame spreads rapidly. (A GHS reference.)

REL Recommended exposure limit.

Reproductive toxin Substances that affect either male or female reproductive systems and may impair the ability to have children.

Respiratory protection Devices that will protect the wearer's respiratory system from overexposure by inhalation to airborne contaminants. Respiratory protection is used when a worker must work in an area where he/she might be exposed to concentration in excess of the allowable exposure limit.

Respiratory sensitizer A substance that induces hypersensitivity of the airways following inhalation of the substance. (A GHS reference.)

Respiratory system The breathing system that includes the lungs and the air passages (trachea or windpipe, larynx, mouth, and nose) to the air outside the body, plus the associated nervous and circulatory supply.

Routes of entry The means by which a material may gain access to the body, for example, inhalation, ingestion, and skin contact.

RTECS The Registry of Toxic Effects of Chemical Substances.

RTK See Right to Know.

Sarcoma A tumor which is often malignant.

Self-Accelerating Decomposition Temperature (SADT) The lowest temperature at which self-accelerating decomposition may occur with substance as packaged. (A GHS reference.)

Self-heating substance A solid or liquid substance, other than a pyrophoric substance, which, by reaction with air and without energy supply, is liable to self-heat; this substance differs from a pyrophoric substance in that it will ignite only when in large amounts and after long periods of time. (A GHS reference.)

Self-reactive substance A thermally unstable liquid or solid substance liable to undergo a strongly exothermic decomposition even without participation of oxygen. This definition excludes substances or mixtures classified under the GHS as explosive, organic peroxides, or as oxidizing. (A GHS reference.)

Sensitizer A chemical that causes a substantial proportion of exposed people or animals to develop an allergic reaction in normal tissue after repeated exposure to the chemical.

SETA Setaflash Closed Tester. See flashpoint.

Signal word A word used to indicate the relative level of severity of hazard and alert the reader to a potential hazard on the label. The signal words used in this section are "danger" and "warning." "Danger" is used for the more severe hazards, while "warning" is used for the less severe.

HAZCOM TERMS & ABBREVIATIONS

Silicosis A disease of the lungs caused by the inhalation of silica dust.

Simple asphyxiant A substance or mixture that displaces oxygen in the ambient atmosphere, and can thus cause oxygen deprivation in those who are exposed, leading to unconsciousness and death.

Skn Skin

Skin A notation which indicates that the stated substance may be absorbed by the skin, mucous membranes, and eyes either airborne or by direct contact and that this additional exposure must be considered part of the total exposure to avoid exceeding the PEL or TLV for that substance.

Skin absorption Ability of some hazardous chemicals to pass directly through the skin and enter the bloodstream.

Skin corrosion Irreversible damage to the skin following the application of a test substance for up to 4 hours.

Skin irritation Reversible damage to the skin following the application of a test substance for up to 4 hours.

Skin lesion An abnormal change in the structure of the surface of the skin due to injury or disease.

Skin sensitizer A chemical that will lead to an allergic response following skin contact.

Solubility The amount of a substance that can be dissolved in a solvent, usually water.

Solution A mixture in which the components are uniformly dispersed. All solutions consist of some kind of a solvent which dissolves the other substance.

Solvent Usually, a liquid in which other substances are dissolved. The most common solvent is water.

Specific gravity The ratio of the weight of a volume of material to the weight of an equal volume of water, usually at 60 ° F, unless otherwise specified.

Spontaneously combustible A material that ignites as a result of retained heat from processing, or which will oxidize to generate heat and ignite, or which absorbs moisture to generate that heat and ignite.

STEL Short-Term Exposure Limit (ACGIH terminology). The airborne concentration of a material to which it is believed that workers can be exposed continuously for a short period of time without suffering from harm. See TLV.

Subcutaneous Beneath the layers of the skin.

Subpart Z Toxic and Hazardous Substances, Tables Z-1, Z-2, and Z-3 of air contaminants, the last subpart of 29 CFR 1910.

Substance Chemical elements and their compounds in the natural state or obtained by any production process, including any additive necessary to preserve the stability of the product and any impurities deriving from the process used, but excluding any solvent which may be separated without affecting the stability of the substance or changing its composition.

Substance which, in contact with water, emits flammable gases A solid or liquid substance or mixture which, by interaction with water, is liable to become spontaneously flammable or to give off flammable gases in dangerous quantities. (A GHS reference.)

Supplemental label element Any additional non-harmonized type of information supplied on the container of a hazardous product that is not required or specified under the GHS. In some cases this information may be required by other component authorities or it may be additional information provided at the discretion of the manufacturer/distributor.

Suspect carcinogen A substance that might cause cancer in humans or animals but has not been so proven.

Symbol A graphical element intended to succinctly convey information. (A GHS reference.)

Systemic poison A poison which spreads throughout the body, affecting all body systems and organs. Its adverse effect is not localized in one spot or area.

Systemic toxicity Adverse effects caused by a substance which affects the body in a general rather than a local manner.

ta Ambient Air Temperature.

Tachycardia Rapid beating of the heart, whether from physiological or pathological causes.

Target organ An organ on which a substance exerts a toxic effect.

Target organ effect Damage caused in a specific organ following exposure to certain chemicals. For example, a "neurotoxin" is a chemical, such as mercury, that produces its primary toxic effect on the nervous system.

Target organ toxin A toxic substance that attacks a specific organ of the body. For example, overexposure to carbon tetrachloride can cause liver damage.

TCC Tagliabue Closed Cup. See flash-point.

TCL Toxic concentration low. Lowest concentration of a gas or vapor capable of producing a refined toxic effect in a specified test species over a specified time.

THA Toxic dose low. Lowest administered dose of a material capable of producing a defined toxic effect in a specific test species.

Technical name A name that is generally used in commerce, regulations, and codes to identify a substance or mixture, other than the YUCCA or SASS name, and that is recognized by the scientific community. (A GHS reference.)

Teratogenic Capable of causing birth defects.

Tfx Toxic effect(s).

Threshold The lowest dose or exposure to a chemical at which a specific effect or reaction is observed.

Tinnitus A sensation of ringing or roaring in the ears caused by a disturbance of the auditory nerve.

TLV (Threshold Limit Value) Threshold Limit Value. The average 8-hour occupational exposure limit. This means that the actual exposure level may sometimes be higher, sometimes lower, but the average must not exceed the TLV. TLVs are calculated to be safe exposures for a working lifetime. This is the ACGIH exposure limit, which will not necessarily correspond to the OSHA-mandated PEL.

TLV-C Ceiling Exposure Limit; the concentration that should not be exceeded even momentarily.

TLV-Skin The skin designation refers to the potential contribution of the overall exposure by the cutaneous route.

TLV-STEL Threshold Limit Value - Short-Term Exposure Limit.

TLV-TWA The allowable Time-Weighted Average concentration for a normal 8-hour workday or 40-hour workweek.

TOC TAG Open Cup. See flashpoint.

Toxic A level of toxicity of a chemical as defined by OSHA in Appendix A of the Hazard Communication Standard, 29 CFR 1910.1200.

Toxicant Any substance producing a toxic effect.

TWA Time Weighted Average.

μ Microgram, one-millionth of a gram.

UN number A four digit number assigned to a material or class of material by the United Nations which are used to identify materials during transportation emergencies.

Uncontrolled release The accidental release of a hazardous substance from its container. If not contained, stopped, and removed, the release would pose a hazard to employees in the immediate area or in areas in the path of the release, or from its by-products or its effects.

HAZCOM TERMS &
ABBREVIATIONS

105

Unstable Tending toward decomposition or other unwanted chemical change during normal handling or storage.

Unstable reactive A chemical that, in the pure state, or as produced or transported, will vigorously polymerize, decompose, condense, or become self-reactive under conditions of shocks, pressure, or temperature.

Upper Explosive Limit (UEL) The maximum concentration of a flammable vapor above which ignition will not occur even on contact with a source of ignition.

Upper Flammable Limit (UFL) The maximum concentration of gas or vapor in air above which it is not possible to ignite the vapors.

Vapor density The density of the gas given off by a substance. It is usually compared with air, which has a vapor density set at 1. If the vapor is more dense than air (greater than 1), it will sink to the ground; if it is less dense than air (less than 1), it will rise.

Vapor pressure The pressure exerted by a vapor, measured in pounds per square inch absolute - psia.

Ventilation See General Exhaust, Local Exhaust, and Mechanical Exhaust.

Viscosity A liquid's internal resistance to flowing.

VOC Volatile Organic Compound. A fast evaporating substance used in coatings and in paints as they evaporate very quickly.

Volatility A measure of how quickly a substance forms vapor at ordinary temperatures.

Water-reactive A chemical that reacts with water to release a gas that is either flammable or presents a health hazard.

Z-List The OSHA table of Permissible Exposure Limits, so named because the tables are identified as Z-1, Z-2, and Z-3, and because they are found in "Subpart Z-Toxic and Hazardous Substances" of the OSHA regulations.

My Workplace Information

Name: _____

Employee #: _____

Department/Work Area: _____

Some locations where hazardous chemicals are used in this workplace are:

_____
_____
_____

Some of the hazardous chemicals that I could be exposed to include:

_____
_____
_____

SDSs for my work area can be found:

_____
_____
_____

Name/title of the person I should ask if I have questions about an SDS or a chemical:

_____
_____

Name/title of the person I should contact if I am injured:

_____
_____

The procedure/number to call in the event of a medical emergency:

_____
_____

The procedure/number to call in the event of a fire:

_____
_____

MY WORKPLACE INFORMATION

My Workplace Information

The procedure/number to call in the event of a chemical spill:

_____

_____

In the event of an evacuation, I need to:

_____

_____

Following an evacuation, my assembly area/head count location is:

_____

_____

Name/title of the person I should ask if I wish to see a copy of the written hazard communication plan:

_____

_____

108

Quiz

Employee _____

Instructor _____

Date _____

Location _____

1. The Globally Harmonized System provides for:

    a. Hazard classifications for chemicals.

    b. A common safety data sheet format.

    c. Common labeling system elements.

    d. All of the above.

2. Chemicals that can cause fire, explosions or some other violent reaction when they come in contact with air, water or other chemicals are known as

    a. Health hazard chemicals.

    b. Environmental hazard chemicals.

    c. Physical hazard chemicals.

    d. Hazard Not Otherwise Classified.

3. The Hazard Communication Standard mandates that your employer must comply with five regulatory requirements: chemical inventory, safety data sheets, labeling, employee training, and a written program.

    a. True

    b. False

109

4. A chemical inventory is:

    a. Supplied by the chemical manufacturer.

    b. A list of hazard information for every chemical used at your company.

    c. Usually comprised of safety data sheets for every chemical used at your company.

    d. A database of all the chemical brand names used at your company.

5. Safety data sheets must have the GHS-specified 16 section format and include certain types of information in each section because:

    a. This standard format complies with the HazCom Standard.

    b. This standard format helps ensure that employees will be able to find safety data sheets in the workplace.

    c. This standard format helps ensure that employers and employees know exactly where to look on a safety data sheet to find information.

    d. Both a and c.

6. According to OSHA, safety data sheets must be readily accessible to you in your work area during each work shift.

    a. True

    b. False

7. HazCom labels include the following element:

    a. Upper and lower flammability limits.

    b. Product identification.

    c. The exact chemical composition.

    d. EPA-mandated disposal methods.

8. A pictogram is:

   a. An image that depicts improper storage methods.

   b. An image that depicts the negative effects of being over-exposed to a chemical.

   c. An image that depicts physical, health, and environmental hazards.

   d. An image that depicts what the chemical is used for.

9. If you want to review your company's written HazCom program, you must first submit a written request to your safety director for approval.

   a. True

   b. False

10. What is an action you should take to stay safe when using chemicals?

    a. Wear proper PPE.

    b. Don't eat around hazardous chemicals.

    c. Wash your hands after using a chemical.

    d. All of the above.

Quiz

Notes